Forces and Motion
with **Wheels**

by Dorothy Heil

Contents

Science Vocabulary

force
A **force** is a push or a pull.

These bikers use **force** to move.

motion
When an object is moving, it is in **motion**.

These wheelchair racers are in **motion**.

lever

A **lever** is a bar or rod used to lift or move things.

These handlebars have **levers** that control the brakes.

ramp

A **ramp** is a flat or curved surface with one end higher than the other.

This biker rides over a **ramp** on a backroads trail.

gravity

Gravity is a force that pulls things toward the center of Earth.

Gravity pulls these skaters down the hill.

attract

To **attract** is to pull toward.

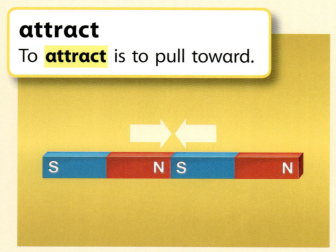

Poles that are different **attract,** or pull together.

repel

To **repel** is to push away.

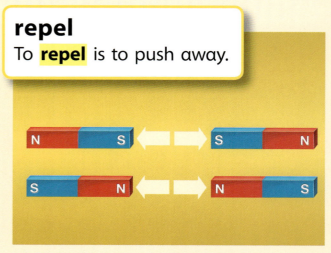

Poles that are the same **repel,** or push away.

magnet

A **magnet** is an object able to pull some metals toward itself.

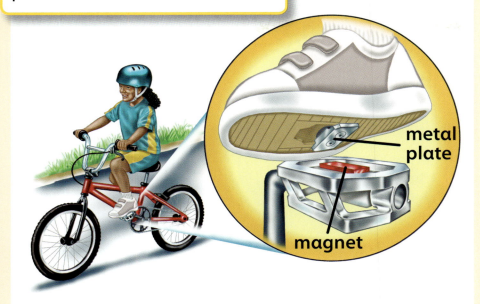

metal plate

magnet

This bicycle pedal has a **magnet** that pulls on a metal piece in the shoe.

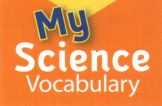

My Science Vocabulary

attract

force

gravity

lever

magnet

motion

pole

ramp

repel

pole

A **pole** is the part of a magnet where its force is the strongest.

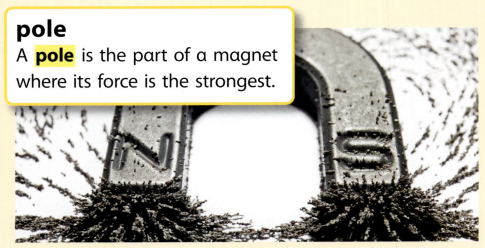

The force of the magnet is strongest at its north **pole** and south **pole.**

7

On Wheels

What would people do without wheels? Wheels help people get to places. People use wheels for many different sports.

Bike riders use **force** to pedal and make the bike wheels turn. A force is a push or a pull. Riders push and pull to move their handlebars and their wheels.

force

A **force** is a push or a pull.

Forces and Motion

People race down the road. They use their arms to push and pull the wheels. Force puts these wheelchairs in **motion**.

The wheelchairs are rolling. They are in motion.

motion

When an object is moving, it is in **motion.**

This person uses force to stop a wheelchair. The person pulls on the wheel to slow it down. The wheel pushes against the ground to come to a stop.

This person has fun skating on a sidewalk. She gives small pushes with her feet. Small pushes help her roll along slowly.

This skater uses a lot of force to race. She pushes hard off the ground to make her wheels turn faster. She uses force to move quickly.

Look at these pictures. People are using wheels in different ways. They use force to put wheels in motion. Some people use pushes. Others use pulls.

Bike

Stroller

Scooter

Wagon

How do these pictures show pushes and pulls?

Push

Pull

push

pull

push

pull

Handlebars on some bikes have **levers,** or straight bars. The levers control the brakes. When the biker pulls the levers, the brake pads press on the wheels.

lever

brakes

lever

A **lever** is a bar or rod used to lift or move things.

The wheels slow down and stop. The force of the wheel pushing against the ground stops the bike.

These bikers ride up and down **ramps**. A ramp is a flat or curved surface with one end higher than the other.

ramp

A **ramp** is a flat or curved surface with one end higher than the other.

This ramp is curved and steep. Riders use a lot of force to go up a steep ramp. Then they roll back down.

The riders use the ramp to go from a high place to a lower place.

Gravity

A girl on a skateboard rolls downhill because of **gravity**. Earth's gravity is a force that pulls down on everything all the time.

gravity

Gravity is a force that pulls things toward the center of Earth.

This skateboarder is soaring through the air.
Gravity will bring him down to the ground.

Gravity keeps these skaters on the ground. It also pulls them down the hill.

Gravity makes it harder to go uphill. People have to use more force to push against gravity. These people would need less force to move on a flat road.

Magnets

Magnets make a pulling force. They attract, or pull, some metals such as iron.

magnet

A **magnet** is an object able to pull some metals toward itself.

attract

To **attract** is to pull toward.

Some bike pedals are magnetic. The magnets help bikers keep their feet on their pedals as they ride on the course.

Magnets on the bike pedals attract the metal plates in the biker's shoes. If the biker falls, the magnet pulls apart from the metal in her shoes.

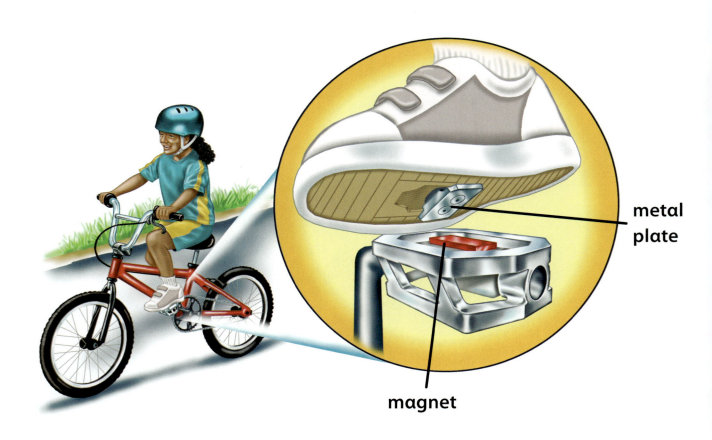

metal plate

magnet

Magnets have two **poles.** Magnets have north and south poles. Poles that are different pull together. Poles that are the same, **repel,** or push away.

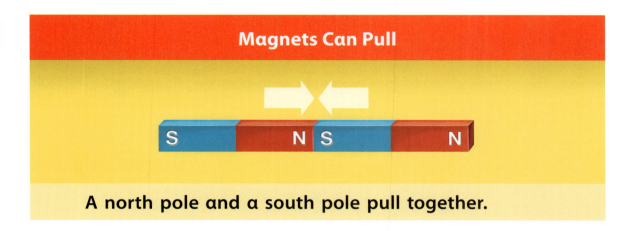

Magnets Can Pull

A north pole and a south pole pull together.

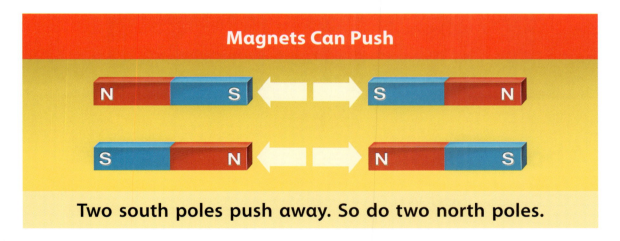

Magnets Can Push

Two south poles push away. So do two north poles.

pole

A **pole** is the part of a magnet where its force is the strongest.

repel

To **repel** is to push away.

Conclusion

People use forces to move on wheels. Some people ride their bikes up ramps. Gravity pulls them back down. Pushes and pulls help bikers move fast. They also help bikers stop. Forces help people get places on wheels.

Think About the Big Ideas

1. How are forces used with wheels?
2. Why is gravity important in skating or skateboarding?
3. How are magnets used on bikes?

Share and Compare

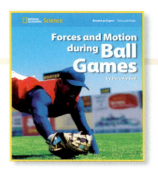  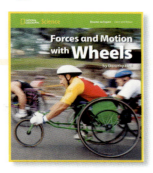

Turn and Talk

Look at the pictures of magnets in your books. Compare the different ways magnets are used.

Read

Find your favorite part of the book and read it to a partner.

Write

Tell about forces in your book. Share what you wrote with a classmate.

Draw

Draw an example of a force. Share your drawing with a classmate.

Meet Marianne Dyson

Marianne Dyson worked for NASA. She helped astronauts plan their days in space. During a space shuttle flight, parts of the shuttle stopped working. The flight had to be cut from five days to two days.

The crew could not do all the tests planned for the flight. To solve the problem, Marianne made a list of the most important tests. The crew used the new plan to finish these tests. Marianne's plan helped make the flight a success.

Index

Acknowledgments
Grateful acknowledgment is given to the authors, artists, photographers, museums, publishers, and agents for permission to reprint copyrighted material. Every effort has been made to secure the appropriate permission. If any omissions have been made or if corrections are required, please contact the Publisher.

Photographic Credits
Cover (bg) Paul Nevin/Photolibrary; Cvr Flap (t), 4 (t), 8-9 imagebroker/Alamy Images; Cvr Flap (c), 6 (t), 22 Scott Markewitz/Getty Images; Cvr Flap (b), 7 (b) Cordelia Molloy/Photo Researchers, Inc.; Title (bg) Paul Paris/Alamy Images; 2-3 Juergen Ritterbach/vario images GmbH & Co.KG/Alamy Images; 4 (b), 10 Philip Scalia/Alamy Images; 5 (t), 16 alejandro Soto/iStockphoto; 5 (b), 18 Matthew Cope Photography; 11 prism_68/Shutterstock; 12 Cory Sorensen/Corbis Premium RF/Alamy Images; 13 John Kelly/Getty Images; 14 (tl), 15 (tr) Timothy Large/Shutterstock; 14 (tr), 15 (tl) Mr_Jamsey/iStockphoto; 14 (bl), 15 (bl) Myrleen Ferguson Cate/PhotoEdit; 14 (br), 15 (br) Sven Klaschik/iStockphoto; 17 Michael Newman/PhotoEdit; 19 Paul Gapper/Alamy Images; 20 Christian Carroll/iStockphoto; 21 Steve Smith/Photodisc/Alamy Images; 23 George Shelley/Corbis; 24-25, 28 Troy Wayrynen/NewSport Photography, Inc.; 31 Bruce Bennett/National Geographic Image Collection; Inside Back Cover (bg) Michelle D. Bridwell/PhotoEdit.

Illustrator Credits
7(t), 26 Gary Torrisi

Neither the Publisher nor the authors shall be liable for any damage that may be caused or sustained or result from conducting any of the activities in this publication without specifically following instructions, undertaking the activities without proper supervision, or failing to comply with the cautions contained herein.

Program Authors
Malcolm B. Butler, Ph.D., Associate Professor of Science Education, University of South Florida, St. Petersburg, Florida; Judith Sweeney Lederman, Ph.D., Director of Teacher Education and Associate Professor of Science Education, Department of Mathematics and Science Education, Illinois Institute of Technology, Chicago, Illinois; Randy Bell, Ph.D., Associate Professor of Science Education, University of Virginia, Charlottesville, Virginia; Kathy Cabe Trundle, Ph.D., Associate Professor of Early Childhood Science Education, The Ohio State University, Columbus, Ohio; Nell K. Duke, Ed.D., Co-Director of the Literacy Achievement Research Center and Professor of Teacher Education and Educational Psychology, Michigan State University, East Lansing, Michigan; David W. Moore, Ph.D., Professor of Education, College of Teacher Education and Leadership, Arizona State University, Tempe, Arizona

The National Geographic Society
John M. Fahey, Jr., President & Chief Executive Officer
Gilbert M. Grosvenor, Chairman of the Board

National Geographic School Publishing
Hampton-Brown
www.NGSP.com

Printed in the USA.
Quad Graphics, Leominster, MA

ISBN: 978-0-7362-5610-0

18 19 20 21

10 9 8 7 6 5 4 3

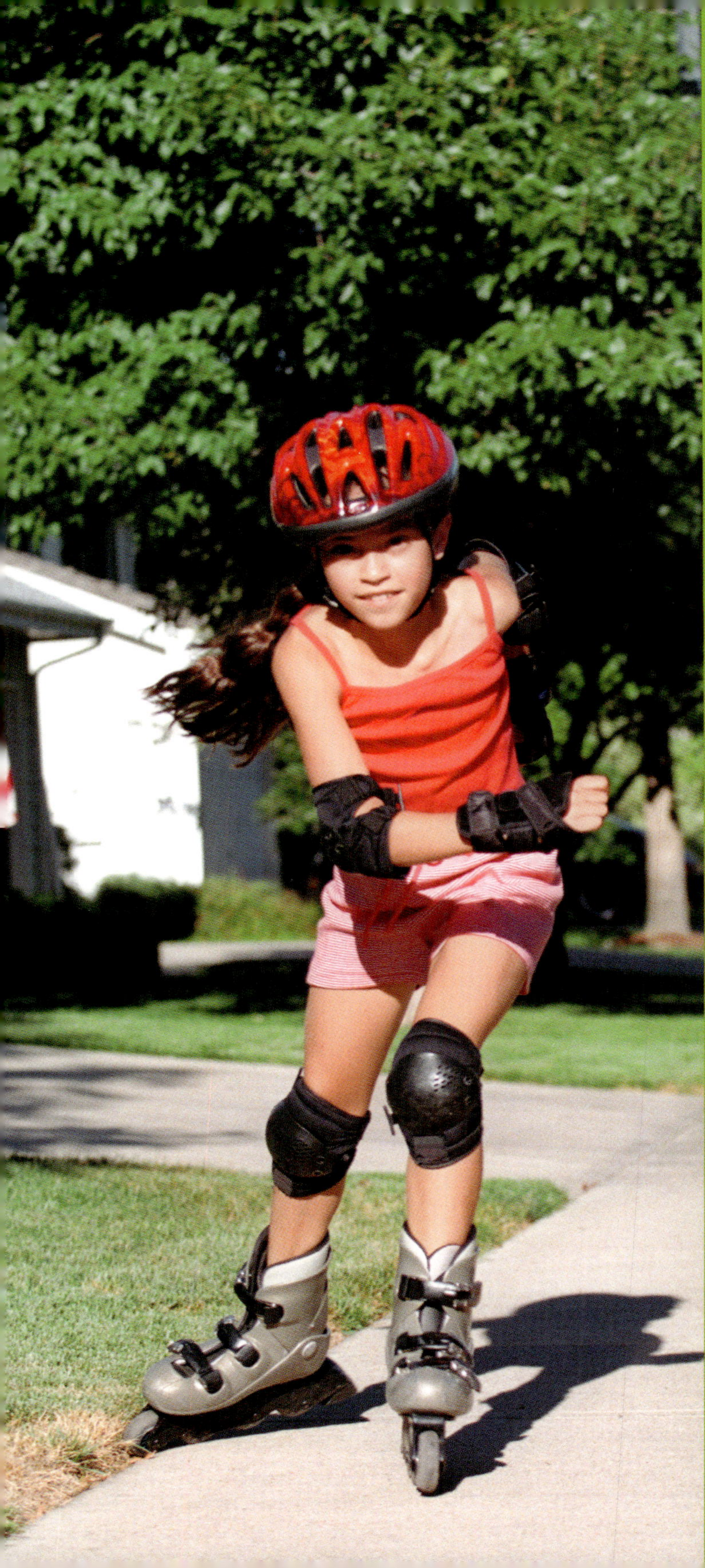

More to Explore

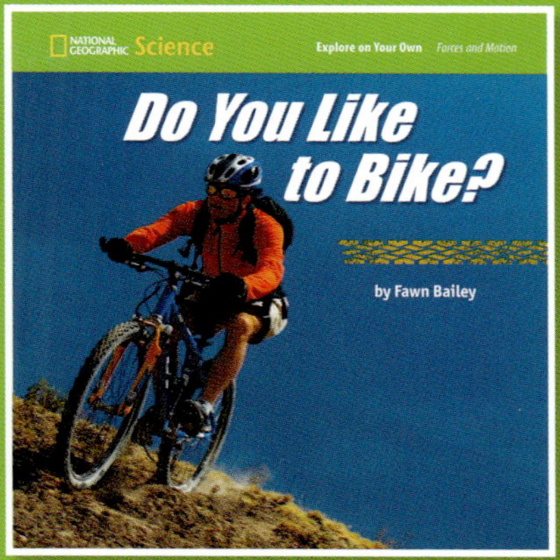

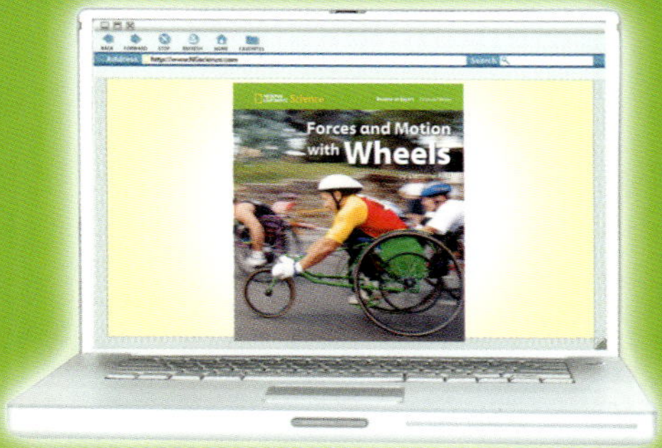

Listen, read, and see pictures come alive at myNGconnect.com

ISBN 978-0-7362-5610-0

9 780736 256100

**NATIONAL
GEOGRAPHIC**

School Publishing